THE OXYGEN ELEMENTS

Oxygen, Sulfur, Selenium, Tellurium, Polonium

Laura La Bella

rosen publishing's

rosen central®

New York

For Paulie

Published in 2010 by The Rosen Publishing Group, Inc.
29 East 21st Street, New York, NY 10010

Library of Congress Cataloging-in-Publication Data

La Bella, Laura.
The oxygen elements: oxygen, sulfur, selenium, tellurium, polonium / Laura La Bella.—1st ed.
    p. cm.—(Understanding the elements of the periodic table)
Includes bibliographical references and index.
ISBN 978-1-4358-3555-9 (library binding)
1. Oxygen—Popular works. 2. Periodic law—Popular works. I. Title.
QD181.O1.L13 2010
546'.72—dc22

2009012541

Manufactured in Malaysia

CPSIA Compliance Information: Batch #TW10YA: For Further Information contact Rosen Publishing, New York, New York at 1-800-237-9932

On the cover: The cover graphic shows the five elements of the oxygen family as they appear on the periodic table. The atomic structure of each element is also shown.

Contents

Introduction

There is one thing we cannot live without: oxygen. It is fundamental to all life on Earth. We know that we can live without food for weeks if we were forced to. We even know that we can live without water for a few days. But life without oxygen . . . it's literally impossible. At most we can last five minutes without oxygen before starting to suffer brain damage and cell death that may be irreversible. This oxygen deprivation can be fatal.

As essential as oxygen is for our bodies, it is also vital to our atmosphere, to Earth's crust, and for the creation of water. Luckily

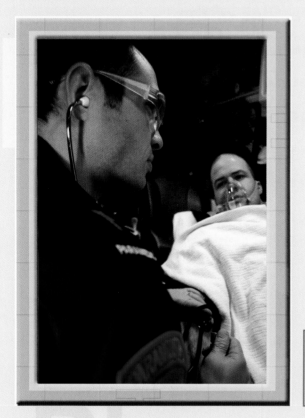

A paramedic treats a patient with oxygen to help him breathe more easily after an emergency.

for us, it is one of the most abundant elements in the world. And the other elements that make up the oxygen family on the periodic table of elements—sulfur, selenium, tellurium, and polonium—each play a crucial role in our lives. They not only aid in our survival, but they also assist in the manufacturing of products that we use each day, helping make our lives easier and more pleasant.

In this book, you will be introduced to the oxygen family and learn how these elements are connected. You will discover how these elements are structured, where they are found, what traits they share, and how each is unique. You will also learn about their uses in everyday life and the many powerful applications that result when they work together and with other elements to form compounds.

Chapter One
Meet the Oxygen Family

Oxygen is all around us. It is in the air we breathe and the water we drink. It is the most abundant element found in Earth's crust, and it is used in the manufacture of products. But we rarely stop to think about how oxygen and the other elements that make up the oxygen family are the foundation for so much life on Earth, and how they make that life possible and sustain it. How does oxygen contribute to the air we breathe and Earth's life-nurturing atmosphere? How does it help form life-giving water? Why is it used so prevalently to make the products that we use every day? To understand how oxygen and the related elements in the oxygen family make all these things possible, we must first understand what these elements are and where they come from.

All About the Oxygen Family

Oxygen is a colorless, odorless, tasteless, gaseous chemical element. Chemical elements can be solids, liquids, or gases. Oxygen is a gas. It is also the third most abundant element in the universe. All chemical elements are organized on a chart called the periodic table of elements. The periodic table is a way for scientists to classify, organize, and compare the more than 100 different elements that exist. Each element has a number, symbol, and specific position on the chart. Oxygen's number on the periodic table is 8, which is called its atomic number. This means that an oxygen atom has eight protons in its nucleus. The element is represented by the symbol O.

Among the most "popular" elements on the periodic table, oxygen can easily form compounds with almost all other elements. A chemical compound is a substance that consists of two or more different elements with their atoms bonded together. For example, water is a compound made up of hydrogen and oxygen. When oxygen atoms bond with hydrogen atoms, water forms as a result.

The periodic table is organized into groups, which consist of elements that all have common characteristics. Oxygen is a member of the

These light and dark blue bubbles are a computer-generated representation of the chemical structure of water molecules. The light blue bubbles are hydrogen atoms, while each dark blue bubble represents an oxygen atom.

chalcogen group, sometimes known as the oxygen family, or group 16. The chalcogen group consists of five elements: oxygen (O), sulfur (S), selenium (Se), tellurium (Te), and polonium (Po). Of these elements, oxygen is the only gas. All of the other chalcogen elements are solids. They all resemble oxygen in some ways, but each is also unique in its own way.

Sulfur

With a chemical symbol of S and an atomic number of 16 (sixteen protons in the nucleus of a sulfur atom), sulfur is a pale yellow, crystalline, nonmetallic solid. It is odorless, tasteless, combustible (able to catch fire), and insoluble (doesn't dissolve) in water. It can react with all metals, except gold and platinum, to form sulfides, which are chemical compounds containing sulfur and one other element. It can also form compounds with several of the nonmetallic elements.

Several million tons of sulfur are produced each year, mostly for the manufacture of sulfuric acid, which is one of the most commonly used chemicals in the world. Sulfuric acid is a combination of sulfur, hydrogen, and oxygen, and it's used in the manufacture of paints, fertilizers, cleaning detergents, synthetic fibers, and rubber.

Selenium

With a chemical symbol of Se and an atomic number of 34, selenium can have either a red or black shapeless structure, or a red or gray crystalline structure. It is rarer than oxygen or sulfur and can occasionally be found in nature accompanying sulfur. More often, however, selenium is found in combination with heavy metals such as copper, mercury, lead, and silver.

Selenium is used in the manufacture of electronic equipment, in pigments, and in glassmaking. It is toxic to animals. Plants grown in selenium-rich soils may have an overabundance of the element that can make the plants poisonous.

This bottle of sulfuric acid comes with a hazard label warning stating that its contents are extremely dangerous. Sulfuric acid is highly corrosive and is used extensively in the chemical industry.

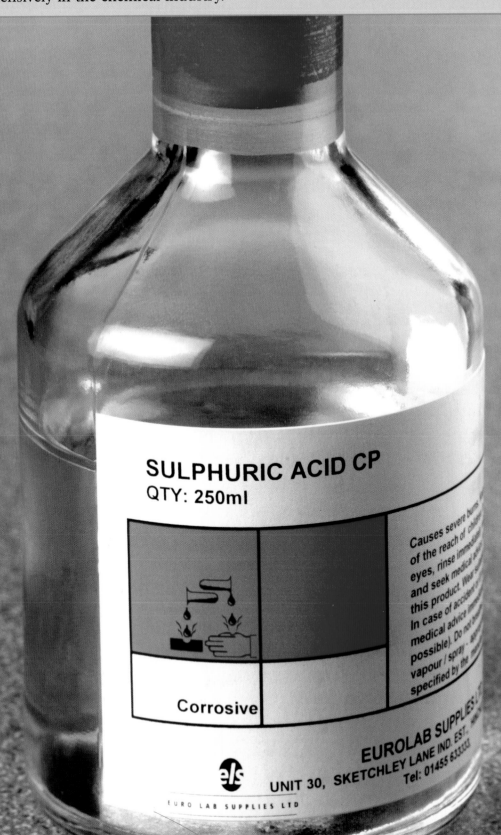

SULPHURIC ACID CP
QTY: 250ml

Causes severe burns.
of the reach of ch
eyes, rinse immed
and seek medical
this product. Wear
In case of accide
medical advice
possible). Do not
vapour / spray -
specified by the

Corrosive

EUROLAB SUPPLIES

UNIT 30, SKETCHLEY LANE IND. EST.
Tel: 01455 633333

els

EURO LAB SUPPLIES LTD

Tellurium

With a chemical symbol of Te and an atomic number of 52, tellurium is silvery white in color and has a metallic luster. Tellurium, like selenium, is very rare and has few actual uses. It is often found as a compound of metals such as copper, lead, silver, and gold, and it is obtained chiefly as a by-product of the refining of copper or lead. No large-scale practical use for tellurium has yet been found.

Polonium

With a chemical symbol of Po and an atomic number of 84, polonium is an extremely rare, radioactive element found in minerals that contain uranium. It has some scientific applications as a source of radiation.

The Difference Between Air and Oxygen

The terms "air" and "oxygen" are sometimes used interchangeably. But the two odorless, colorless gases are really quite different from one another. Oxygen is a pure element, and exposure to it in its pure form can be very harmful. By contrast, air is a healthy mixture of elements that sustain human life (except when air pollution makes it toxic or harmful to breathe).

The air we breathe is primarily a mixture of oxygen and nitrogen. Nitrogen generally makes up approximately 78 percent of the air we breathe, while oxygen makes up 21 percent. Other gases—including carbon dioxide, hydrogen, helium, argon, and neon—make up the remainder. The term "air" is also used to refer to Earth's atmosphere as a whole: a delicate and unique balance of gases that has made life on Earth possible.

Antoine-Laurent Lavoisier was one of three scientists in the 1700s who first discovered the element oxygen. Lavoisier is also credited with giving the element its name.

The Discovery of Oxygen

The credit for the discovery of oxygen is commonly split among three scientists: Joseph Priestley (1733–1804), Carl Wilhelm Scheele (1742–1786), and Antoine-Laurent Lavoisier (1743–1794). They discovered the element independently of each other. Scheele is thought by many to be the first of the three to make the discovery, in Uppsala, Sweden, in the early 1770s. Priestley made the discovery in Wiltshire, England, in 1774. But it was Lavoisier, a French chemist, who not only helped to discover the element but is also credited with naming it "oxygen." (The word "oxygen" is a combination of Greek words that means "begetter of acids," since Lavoisier mistakenly believed that oxygen helped form acids.) Priestley is often given priority as the element's chief discoverer, but only because his research was published before Scheele and Lavoisier could report their findings.

An Essential and Useful Element

No matter who is given credit for the discovery of oxygen and its many uses, it is among the most important finds in science. Oxygen is vital

Chloroplasts, found in the cells of plants, conduct photosynthesis, which converts carbon dioxide into organic compounds like sugar. In the process, oxygen is released into the air.

for all life on Earth. It is a central component of a crucially important biological function that is necessary for all people, animals, and plants to survive: the oxygen cycle. (For a more thorough discussion of this, see chapter 3.)

We receive our oxygen courtesy of plants and trees. In the leaves of plants and trees, photosynthesis converts the carbon dioxide in the air into organic compounds, especially sugars, using sunlight. When this conversion takes place, oxygen is released into the atmosphere by the vegetation as a by-product of photosynthesis. In more simple terms, plants "breathe" in carbon dioxide and "breathe" out oxygen. Animals and humans breathe in oxygen and breathe out carbon dioxide. Together, the oxygen cycle and photosynthesis are responsible for the creation of Earth's atmosphere and the oxygen we breathe and require to survive.

In addition to its vital role in sustaining life on Earth, oxygen has also proven helpful in treating illnesses such as emphysema, pneumonia, and some heart disorders. It can be a useful treatment for diseases that impair the body's ability to take in and use oxygen. In hospitals, oxygen is used to enhance breathing or respiration. It is also used in situations in which oxygen levels are low or impossible to access. For example, oxygen is delivered via modern space suits for astronauts and through masks or mouthpieces for scuba divers and pilots flying at high altitudes.

Most modern industries use oxygen and other chalcogen elements to manufacture products. For example, the smelting of iron ore into steel consumes 55 percent of the oxygen that is commercially produced. Selenium, a member of the oxygen family, is vital to the imaging and photographic industry. It is used as the light-sensitive component in photocopiers. Because it conducts electricity when exposed to light, selenium is also used in solar cells, light meters, and photographic materials.

Chapter Two
The Oxygen Family's Place on the Periodic Table

There are currently 118 elements on the periodic table. It's important to understand how these elements are classified and organized. The periodic table classifies and compares the many different forms of chemical behavior exhibited by each element. Each element is different in its structure, and each has its own characteristics or properties that make it unique.

Each Element Is Unique

Elements are distinctive in their own ways, and no two are exactly alike. Elements are different because their atoms are different. All elements are made up of atoms, which contain protons, electrons, and neutrons. Neutrons and protons are found in the central, or middle, part of the atom. This central part is called the nucleus. Electrons are found outside and around the nucleus.

Protons, electrons, and neutrons are all particles of matter, but each has a different electric charge. Protons have a positive charge. Electrons have a negative charge. Neutrons have no charge at all. There is always an equal number of electrons and protons in a neutral atom. All elements on the periodic table are assigned an atomic number. This is the number

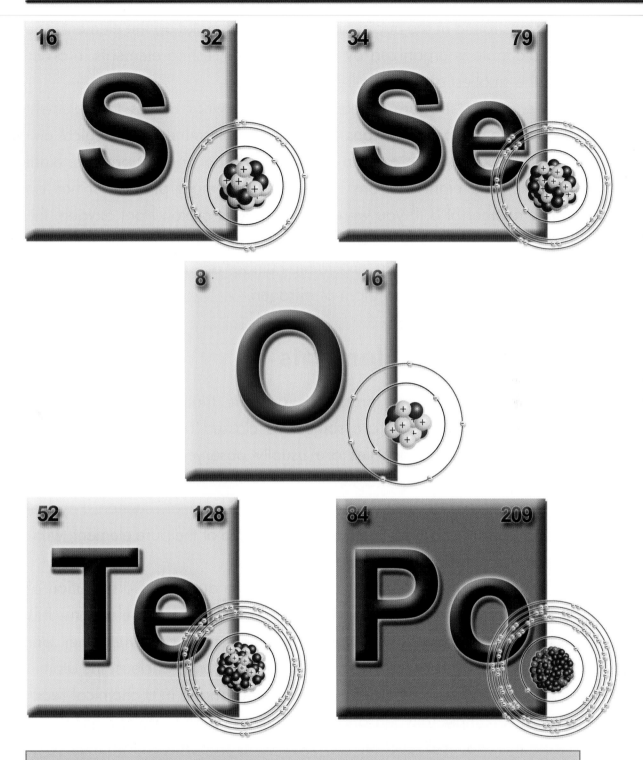

The elements of the oxygen family (sulfur, selenium, oxygen, tellurium, and polonium) are shown here as they appear on the periodic table of elements.

of protons found in the nuclei of the atoms that make up the element. The periodic table is organized by atomic number, listing elements in order from the smallest atomic number to the largest.

The atomic structure of each element is what makes that element unique. The number of protons in the atoms determines the chemical element. For example, an oxygen atom is made up of eight protons, eight electrons, and eight neutrons. Because it has eight protons, it has an atomic number of 8. If you were to remove one proton from oxygen, the element would change. It would become nitrogen—another colorless, odorless gas. Nitrogen has an atomic number of 7, indicating that there are seven protons in each atom of nitrogen.

Properties of Elements

Each element has its own properties, or traits, that make it unique. The properties of an element are sometimes classed as either chemical or physical. Chemical properties are usually observed when a substance interacts with other substances and forms new substances. The interaction and formation of new substances is called a chemical reaction. Physical properties are observed by examining a sample of the pure element, without changing it into something else.

The chemical properties of an element are due to the distribution of electrons around the atom's nucleus, particularly the outer electrons. It is these electrons that are involved in chemical reactions with other elements. A chemical reaction does not affect the atomic nucleus; the atomic number of an element therefore remains unchanged in a chemical reaction. For example, oxygen and hydrogen combine to create water. But during this chemical reaction, oxygen and hydrogen maintain their original number of protons. The oxygen atoms do not change, nor do the hydrogen atoms. The oxygen atoms in water still have eight protons, and

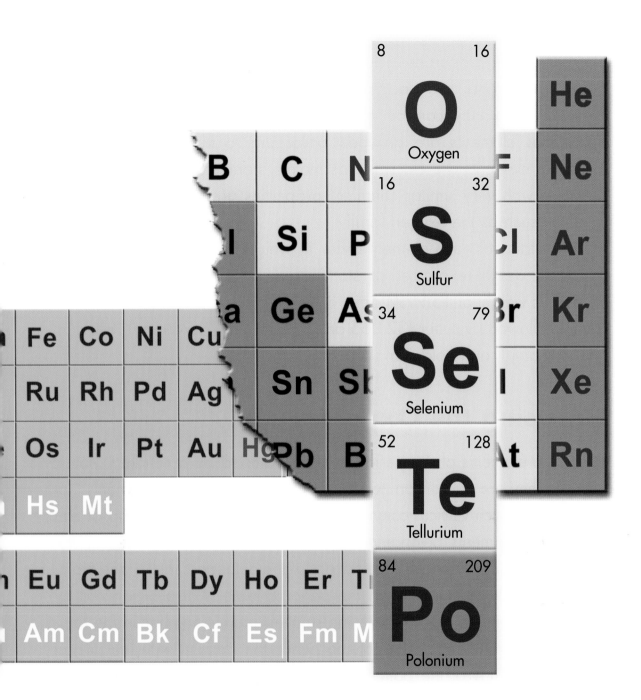

The periodic table of elements is organized by families, the members of which all have common traits. Each vertical column represents a family. Shown here is the oxygen family's place on the periodic table.

the hydrogen atoms have one proton. Some properties of an element can be observed by looking at a collection of atoms or molecules of that element. These properties include color, density, melting point, boiling point, and thermal and electrical conductivity. These are physical properties because they can be observed without changing the element. Sometimes elements are grouped according to their properties. For example, elements are classified as metals, nonmetals, and metalloids. Elements with very similar chemical properties are often referred to as families, such as the oxygen family.

The atoms of each element differ in weight, and each is measured by atomic weight (expressed as atomic mass units, or amu). Oxygen has an atomic weight of 15.9994 amu. It is the lightest element of the oxygen family, followed by sulfur (32.065 amu), selenium (78.96 amu), tellurium (127.60 amu), and polonium (209 amu).

Grouping the Oxygen Family

While the elements of the periodic table are listed by ascending atomic number, they are also organized into groups, or families, of elements that share traits. Each vertical column of the table represents a group of elements. The elements belonging to group 6—the oxygen family or the chalcogen group—all have six electrons in the outermost shell of the atom.

Each family of elements on the periodic table is also organized by atomic number. The element with the smallest atomic number in the family appears at the top of the column, while the element with the largest atomic number is at the bottom. In the oxygen family, oxygen has the smallest atomic number of its group, followed in ascending order by sulfur, selenium, tellurium, and polonium.

The elements in families in the periodic table become more similar down each column. The element at the top of each column is less like the

Snapshot of the Oxygen Family

Name	Symbol	Atomic number	Phase	Element category	Appearance
Oxygen	O	8	Gas	Nonmetal	Colorless gas
Sulfur	S	16	Solid	Nonmetal	Yellow crystals
Selenium	Se	34	Solid	Nonmetal	Gray-black, metallic luster
Tellurium	Te	52	Solid	Metalloid	Silvery, lustrous gray
Polonium	Po	84	Solid	Metalloid	Silvery

others. In the oxygen family, oxygen stands out as the least like the others. It is a gas, while the others are solids. It forms compounds with most of the other elements in the periodic table, while the other members of the family combine with only some other elements.

Oxygen is extremely active chemically, forming compounds with almost all the elements, except the inert gases. In addition, oxygen can combine with a number of other elements to form oxides. An oxide is a chemical compound that contains at least one oxygen atom, as well as at least one other element. For this reason, oxygen is also an important component of carbohydrates, proteins, fats and oils, alcohols, and numerous other compounds like the carbonates, chlorates, nitrates, phosphates, and sulphates.

Chapter Three
Basic Profiles of the Oxygen Elements

Oxygen is a vital element in our atmosphere, and it's a crucial component of water. The major nutrients that we need to survive—proteins, carbohydrates, and fats—all contain oxygen, as do the major inorganic compounds that make up animal shells, teeth, and bone. Oxygen may be the most abundant and readily available element of the oxygen group, but the remaining chalcogen elements are also all around us, though sometimes in less accessible places.

Oxygen

Of the 118 elements on the periodic table, 92 elements occur naturally, which means that they can be found in nature and do not have to be created in a laboratory. Oxygen is one of them. Oxygen is abundant in Earth's rocks and minerals. And because it readily combines with other elements, it can form any number of compounds.

Oxygen is the most abundant, by mass, in the air, sea, and land. Oxygen is the third most abundant chemical element in the universe, after hydrogen and helium. About 0.9 percent of the sun's mass is contributed by oxygen, and the element constitutes more than 46 percent of Earth's crust. Because water is formed by the combination of oxygen

Oxygen is a major component of the water that makes up lakes, rivers, and oceans. It is also the second most common element found in Earth's atmosphere.

and hydrogen, oxygen is the major component of the world's oceans, rivers, streams, ponds, and lakes. Oxygen gas is the second most common component of Earth's atmosphere, contributing about 21 percent of its volume. This unusually high concentration of oxygen gas makes Earth unique among the planets of our solar system. Earth is the only planet that has been found to be able to sustain complex life forms, such as animals and humans. This ability to generate and sustain life is the result of the oxygen cycle.

The oxygen cycle is a complex series of processes in which oxygen atoms present on Earth and in its atmosphere circulate between Earth's atmosphere, biosphere, and lithosphere. The driving force of the oxygen cycle is photosynthesis, which is responsible for the creation and maintenance of Earth's atmosphere. Photosynthesis occurs in plants, which use energy from sunlight to convert carbon dioxide and water into carbohydrates and oxygen. This means that plants "breathe" in carbon dioxide and "breathe" out oxygen, releasing it into the atmosphere.

Animals (including humans) play an equally important role in the oxygen cycle. In fact, they form the other half of the cycle. They breathe

Humans and animals take in oxygen and breathe out carbon dioxide. The visible breath of this horse shows carbon dioxide being expelled into the air.

in oxygen, which they use to break carbohydrates down into carbon dioxide and release energy, in a process called respiration. Carbon dioxide produced during respiration is breathed out by animals and released into the air.

While oxygen is, literally, all around us, it is never found floating around as an individual atom. It is always combined with another atom, even if the "other" atom is another oxygen atom. When two oxygen atoms combine, they form diatomic molecules, O_2, sometimes called dioxygen. Diatomic oxygen gas makes up approximately 20 percent of the volume of air. When three oxygen atoms combine, they form trioxygen (O_3), or what's commonly referred to as ozone. Ozone is found in two different places in Earth's atmosphere—in the ozone layer high in the atmosphere and low-level ozone near the surface of Earth.

The ozone layer filters out the sun's potentially damaging ultraviolet light and keeps it from reaching Earth's surface. This helps shield living organisms, including humans, from the harmful radiation from the sun. Low-level ozone is regarded as a pollutant by the World Health Organization (WHO) and the U.S. Environmental Protection

Pollution, seen here as heavy smog on this highway in Beijing, China, depletes oxygen from our atmosphere and damages Earth's ozone layer.

Agency (EPA). It is formed by the reaction of sunlight on air containing hydrocarbons and nitrogen oxides. This results in smog forming at the pollution site. High levels of smog are usually found in large cities, such as Los Angeles and Houston, where there are many vehicles, because vehicles can emit hydrocarbons and nitrogen oxides.

Sulfur

Sulfur is a very common element that is found as a yellow crystalline solid. In nature, it is found as both the pure element and as sulfide and sulfate minerals. It is an essential element for life.

Sulfur is used in a number of different ways. One of its most popular uses is in the creation of fertilizer when it is combined with phosphate. Farmers add fertilizer to soil to help plants—mainly fruits and vegetables—grow.

Sulfur has a wide range of uses. It is a component of gunpowder and is used to harden rubber. Sulfur is commonly used in the production of fertilizer, as well as sulfuric acid, which is a widely used industrial chemical. Some sulfur compounds can be highly toxic. For example, small amounts of hydrogen sulfide can be metabolized, but higher concentrations can quickly cause death from respiratory paralysis. In addition, hydrogen sulfide quickly deadens one's sense of smell.

Polonium

An extremely dangerous element, polonium is very rare because of the short-lived nature of its isotopes. The atoms of the various isotopes of an element all contain the same number of protons, but they have different numbers of neutrons. Because polonium is a very unstable element, it requires specialized equipment and strict handling procedures to ensure safety. How dangerous is it? One gram (1,000 milligrams) of the element will self-heat to a temperature of around 932 degrees Fahrenheit (500 degrees Celsius).

Polonium is used to eliminate static charges in places like textile mills, where sparks can create devastating fires. And it can be used as an atomic heat source to power thermoelectric generators. Because of its very high toxicity, polonium is sometimes used as a poison (in fact, polonium is found in tobacco smoke). Most often, however, the element is used for getting rid of dust on motion picture and photographic film.

Selenium

Selenium rarely occurs in its elemental state in nature. When needed, it can be produced from its compounds, most commonly from selenide, which is found in sulfide ores like copper, silver, or lead. It is obtained as

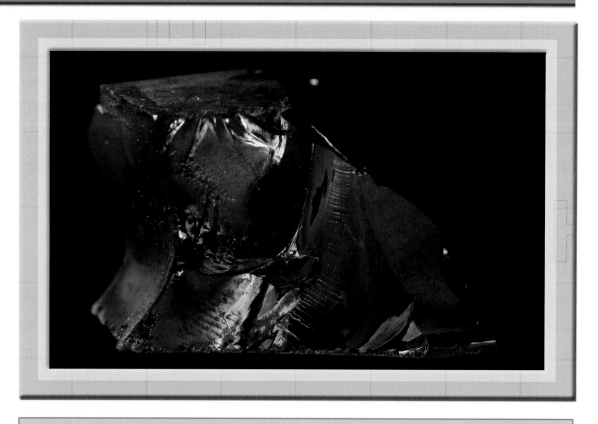

This chunk of hard, gray-black rock is a sample of selenium crystal. Selenium is necessary for healthy cellular function in the human body. However, large doses can be extremely toxic.

a by-product of the processing of these ores. Selenium is toxic to humans in large amounts and can lead to a condition called selenosis. Symptoms of selenosis include a garlic odor on the breath, gastrointestinal disorders, hair loss, falling off of fingernails, fatigue, irritability, and neurological damage. Yet small amounts of selenium are necessary for healthy cellular function in most, if not all, animals.

Selenium is most commonly used in manufacturing. It is an essential material in the light-sensitive drums of laser printers and copiers. The largest use of selenium worldwide is in glass and ceramic manufacturing, where it is used to give a red color to glass, enamel, and glazes. It is also

Oxidation

A change occurs when oxygen combines with certain materials or compounds. This change is called oxidation. A freshly cut apple turns brown. A bicycle fender becomes rusty. A shiny new copper penny slowly darkens and becomes dull. These are all examples of oxidation. The exposure to oxygen changes the appearance of these items.

The process of oxidation depends on the nature of the material it touches. Oxidation happens on a molecular level. In the case of fresh fruit, its skin usually provides protection against oxidation. This is why most fruits and vegetables arrive in good condition at the grocery store. Once the skin has been broken, however, the individual cells of the fruit come into direct contact with the air, and the oxygen molecules in the air start burning them. The result is brownish spots or blemishes. Eventually, the fruit or vegetable spoils.

used to hide unwanted color from glass by counteracting the green tint that is caused by impurities.

Tellurium

A brittle, silvery white metalloid that looks a lot like tin, tellurium is chemically related to selenium and sulfur. It is primarily used in alloys and as a semiconductor. Tellurium is extremely rare. In fact, it is one of the nine most rare metallic elements on Earth (excluding radioactive metals). Tellurium is often added to lead to improve its strength and durability.

Chapter Four
Compounds of the Oxygen Elements

Oxygen is a highly reactive element that very easily combines with almost every element found on the periodic table. A compound is a substance that forms when two or more elements are chemically joined. Water, salt, and sugar are all examples of compounds. When the elements are joined in a compound, the elements lose their individual properties and the compound has properties different than those of the elements it contains.

Oxygen combines with all other elements except the noble gases helium, neon, and argon. These are elements whose outer shells of electrons are considered to be "full," allowing them little to no opportunity to bind with other elements to create compounds. Oxygen even combines with some of the most rare elements, including technetium (technetium dioxide; TcO_2), promethium (promethium oxide; Pm_2O_3), and neptunium (neptunium dioxide; NpO_2). It can also combine with some of the least reactive elements, such as xenon (xenon trioxide; XeO_3), gold (gold oxide; Au_2O_3), and platinum (platinum oxide; PtO_2).

Common Oxygen Compounds

Some of the most common and most important compounds in everyday life are compounds of oxygen with another element.

Water (H_2O) is perhaps the most common and familiar of all oxygen compounds. Water is an oxide. An oxide is a chemical compound containing oxygen combined with another element. Water is a compound of oxygen with hydrogen, and each molecule of water contains two hydrogen atoms plus one oxygen atom. As we all know, water is essential for the survival of all forms of life. The term "water" usually refers only to its liquid form or state. But the substance also has a solid state, which occurs when the temperature goes down to its freezing point or below, and water becomes ice. Water's gaseous state, often referred to as water vapor or steam, is present in the air around us. When liquid water is heated to its boiling temperature, all of the liquid will turn into vapor—the result of

This photo shows what oil looks like when mixed with water. Because oil is one of the substances that water cannot dissolve, the oil sits on top of the water.

water reaching its boiling point. Water is the only substance that is present on Earth in all three phases, solid, liquid, and vapor.

Water is a very important solvent. This means that it can form a solution with other substances. When a substance dissolves in water, a clear solution is formed, one in which every drop is exactly alike. Substances that dissolve in water include many salts, sugars, and acids. Substances that dissolve in water are known as hydrophilic, or water-loving, substances. Those that do not dissolve in water, like oils and fats, are known as hydrophobic, or water-fearing, substances.

Another common oxide is carbon dioxide (CO_2). Its molecules are composed of two oxygen atoms that are bonded to a single carbon atom. Plants use it during photosynthesis. Carbon dioxide is also the waste product of human respiration, or what we breathe out through our noses and mouths. The compound is also a greenhouse gas. Greenhouse gases are essential to determining the temperature of Earth because they help trap heat near Earth's surface, allowing life to flourish. Without them this planet would likely be too cold to be inhabitable. Yet with the current buildup of greenhouse gases like carbon dioxide in Earth's atmosphere (due to things like car exhaust, factory emissions, and fossil fuel-burning),

An Oxygen That Can Help Save the Environment

Best known as the type of alcohol that is found in alcoholic beverages and modern thermometers, ethanol (C_2H_6O) is also becoming a popular fuel for cars and trucks. This hydrogen-oxygen-carbon compound is a volatile, flammable, colorless liquid that can be used as a more Earth-friendly motor fuel and fuel additive, since it is made from grain, which is a renewable resource unlike carbon-based fossil fuels.

the planet's surface temperatures are rising, perhaps eventually to a dangerous and life-threatening level.

Carbon dioxide gas can be found mainly in air. The most familiar example of its use is in soft drinks, to make them fizzy. Carbon dioxide is released by baking powder or yeast to make cake batter and dough rise. It is also used in some fire extinguishers because it does not support combustion, and it is denser than air and can blanket a fire. As a result, the fire, which needs oxygen to keep burning, is extinguished. Carbon dioxide can become solid only when temperatures drop below -109.3°F (-78.5°C).

Carbon monoxide (CO) is a colorless, odorless, tasteless gas. Its molecules consist of one carbon atom that is bonded to one oxygen atom. Carbon monoxide is a highly toxic gas and can lead to fatal poisoning if inhaled. The gas is especially dangerous to humans because it is not detected by our sense of smell.

Today, carbon monoxide is considered a pollutant, even though it has always been present in the atmosphere. It occurs naturally as a product of volcanic activity, but two-thirds of carbon monoxide emissions come from vehicles, mostly cars. In urban areas, transportation sources account for as much as 95 percent of an area's carbon monoxide pollution. Even though it can cause pollution, carbon monoxide is used in the manufacturing of products and is sold as a major industrial gas. Industrial gases are mainly used in steelmaking, oil refining, medical applications, fertilizer, and semiconductors.

Calcium carbonate ($CaCO_3$) is found throughout the world in the form of limestone rock, and the compound is the main component in the shells of marine organisms, snails, and eggshells. It is a compound of oxygen with calcium and carbon. Calcium carbonate is only very slightly soluble in water. "Hard water" usually has a small amount of calcium carbonate dissolved in it. Calcium carbonate is used in making agricultural lime. It also has medicinal qualities. Calcium carbonate is often used as a calcium supplement or

Hydrogen peroxide, a weak acid, is commonly used to clean cuts and wounds. The chemical, found in liquid form, is easily purchased at drugstores and supermarkets.

as an antacid. However, overconsumption of it can be hazardous.

Hydrogen peroxide (H_2O_2) is another oxide of hydrogen. Unlike water, its molecules consist of two oxygen atoms bonded to two hydrogen atoms. It is a very weak acid that is familiar to most people as an over-the-counter liquid used to clean minor wounds and bleach hair blond. It is easily available at supermarkets as well as pharmacies. The compound is also widely used by paper manufacturers to bleach or dye paper white.

Oxygen forms several compounds with iron. They differ in the ratio of iron atoms to oxygen atoms. One of them is iron oxide (Fe_2O_3), or rust, in which there are three oxygen atoms for every two iron atoms. Rust is a general term for iron oxides formed by the reaction of iron and oxygen in the presence of water or air moisture. Rust can be prevented by coating metals, such as iron and steel, to keep water or moisture from coming into contact with them. Another iron oxide contains the same number of oxygen atoms as iron atoms. Its chemical formula is FeO, and it is found in some minerals.

Nitric oxide (NO) is a substance whose molecules contain one atom of nitrogen and one atom of oxygen. When pure, nitric oxide is a poisonous gas that is used in making nitric acid, an important industrial chemical. It is also formed inside automobile engines, and unless it is

Industrial plants often produce sulfur dioxide as part of the manufacturing process. Here, a large plume of exhaust is blown into the sky, creating pollution and damaging the atmosphere.

removed from the exhaust, it is a major form of air pollution. However, nitric oxide is also an important biological molecule, where it carries signals from one cell to another.

Sulfur dioxide (SO_2) is a gas produced by volcanoes and by various industrial processes. Its molecules contain two atoms of oxygen attached to one atom of sulfur. Sometimes sulfur dioxide is used as a preservative for dried apricots and other dried fruits. It maintains the appearance of the fruit and prevents it from rotting. Sulfur dioxide is also a very important preservative in winemaking. It serves as both an antibiotic and an antioxidant, protecting wine from bacteria and oxidation, which can cause the wine to spoil.

Chapter Five
The Oxygen Elements and You

Nearly every day, we interact with elements of the oxygen family without giving much thought to how they impact our life. Oxygen itself has the most obvious of practical uses. In addition to providing us with life-giving air and water, oxygen helps burn fossil fuels, is essential for welding and steelmaking, and has a number of medical applications. Yet the other elements of the oxygen family all have practical uses as well, some of which may surprise you.

Oxygen

Oxygen is widely used as an aid in breathing. Astronauts and scuba divers use oxygen supplies when they enter environments

A NASA astronaut is able to float free of the space shuttle and perform a spacewalk, thanks to the oxygen stored in his spacesuit.

(outer space and underwater) where the element is limited, nonexistent, or not in a breathable form. People who climb mountains or fly at high altitudes may also have supplemental oxygen supplies with them. Passengers traveling in pressurized commercial airplanes are automatically provided with an emergency supply of oxygen in case of cabin depressurization.

Oxygen has medical and therapeutic uses. A patient having difficulty breathing is given doses of pure, or nearly pure, oxygen. This type of therapy is used to treat emphysema, pneumonia, some heart disorders, and any disease that impairs the body's ability to use gaseous oxygen.

Ozone, a different molecular form of elemental oxygen, is used to clean aquariums, swimming pools, and spas. It is also used to maintain a healthy environment in ponds. Some cities, such as Paris, use ozone to sterilize drinking water. Ozone minimizes bacterial growth, controls parasites, eliminates the transmission of some diseases between fish, and reduces or eliminates the discoloration of water. It is often used to disinfect laundry in hospitals and clean food processing plants.

In the processing of iron ore in a blast furnace, oxygen is used to convert coke (carbon) to carbon monoxide. The carbon monoxide, in turn, reduces the iron oxides to pure iron metal.

Oxygen is used in rockets and missiles. The space shuttle, for instance, carries huge internal tanks

When oxygen and hydrogen interact in the space shuttle's fuel tank, the resulting combustion provides a powerful lift that thrusts it into outer space.

containing liquid oxygen and liquid hydrogen. These two are vaporized, mixed, and ignited. When they burn to form water vapor, the resulting combustion gives the vehicle enormous thrusting power that helps lift it into space.

Sulfur

One of the most common uses for sulfur is in the manufacture of rubber. Sulfur is used to harden rubber, making it more durable and resistant to breakage and cracks.

Gunpowder, also called black powder, is an explosive mixture of sulfur, charcoal, and potassium nitrate. It burns rapidly, producing hot gases that can be used as a propellant in firearms and fireworks.

These gunpowder kegs, left over from the Civil War, are full of an explosive mixture of sulfur, charcoal, and potassium nitrate.

Sulfur is the only fungicide that is used in organically farmed apple production. It helps fight against diseases that can damage the apple crops. It is also a major fungicide in the farming of grapes, strawberries, and many other crops. Sulfur is also used as a natural insecticide that is known to be effective against mites.

Selenium

Selenium is a catalyst that increases the speed of many chemical reactions. It is used in the manufacture of glass and ceramic. Selenium gives a red color to glass, enamel, and glazes, and it removes unwanted color from glass.

Oxygen Bars: Healthy or Hype?

Peppermint. Bayberry. Cranberry. Wintergreen. No, these aren't varieties of scented candles. They are flavors of oxygen that are available at your local oxygen bar. Oxygen bars were introduced to the United States in the late 1990s and instantly became popular. Customers pay to inhale oxygen through a plastic hose for up to twenty minutes. But does this trend have any real impact on a person's overall health?

Most oxygen bar customers claim that inhaling oxygen can reduce stress, increase energy and alertness, lessen the effects of headaches and sinus problems, and generally be relaxing for the body. However, the American Lung Association says that inhaling oxygen at these bars is unlikely to have a beneficial physiological effect, and, in some instances, it can be dangerous if a customer has an unknown medical condition, such as asthma, heart problems, or certain lung diseases.

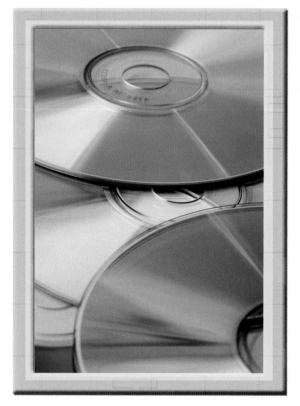

Selenium is widely used in electronic devices, from cameras to CDs and DVDs. It is used in light meters and photo cells, and it can even extend the life of photographs.

Selenium is an unusual material that becomes electrically conductive when exposed to bright light. Because of this property, it is used in photocopiers, photocells, light meters, and solar cells. Because it is a semiconductor, selenium is widely used in electronic devices. Selenium is used to tone photographic prints. Its use intensifies and extends the range of black-and-white photographs and improves the life of a photographic print.

Tellurium

Tellurium is mostly used in alloys with other metals. It is added to lead to improve its strength and durability, and to decrease the corrosiveness of sulfuric acid. When added to stainless steel and copper, tellurium makes these metals more workable.

As the compound tellurium suboxide, tellurium is used as a layer on several types of rewritable CDs, DVDs, and Blu-ray discs. It is also used in the memory chips that are developed by Intel, a leading manufacturer of microchips, processors, and other computer and communications products.

Polonium

Polonium is used in the brushes that remove dust from photographic film. It is sealed in the bristles of these brushes and releases radiation, which creates an electric charge. This electricity, in turn, neutralizes the static electricity that can build up on the film's surface and result in poor picture clarity.

Polonium has been used as a lightweight heat source to power thermo-electric cells in artificial satellites. A polonium heat source was used in the land rovers that were deployed on the surface of the moon to keep the vehicles' internal components warm during frigid lunar nights.

As we have seen, the elements of the oxygen family are essential to the fostering and sustaining of plant and animal life. And these elements and the compounds they form are highly useful to making human life far more pleasant and comfortable, and increasingly high-tech. Without the oxygen family, the human family would not be where it is today. In fact, it wouldn't be here at all!

The Periodic Table of Elements

Group

IA	IIA	IIIB	IVB	VB	VIB	VIIB	VIIIB	VIIIB
1	2	3	4	5	6	7	8	9

Atomic Number →

8 · O · Oxygen · 16	16 · S · Sulfur · 32	34 · Se · Selenium · 79

Name of Element

Period

Period	IA (1)	IIA (2)	IIIB (3)	IVB (4)	VB (5)	VIB (6)	VIIB (7)	VIIIB (8)	VIIIB (9)
1	1 **H** 1 Hydrogen								
2	3 **Li** 7 Lithium	4 **Be** 9 Beryllium							
3	11 **Na** 23 Sodium	12 **Mg** 24 Magnesium							
4	19 **K** 39 Potassium	20 **Ca** 40 Calcium	21 **Sc** 45 Scandium	22 **Ti** 48 Titanium	23 **V** 51 Vanadium	24 **Cr** 52 Chromium	25 **Mn** 55 Manganese	26 **Fe** 56 Iron	27 **Co** 59 Cobalt
5	37 **Rb** 85 Rubidium	38 **Sr** 88 Strontium	39 **Y** 89 Yttrium	40 **Zr** 91 Zirconium	41 **Nb** 93 Niobium	42 **Mo** 96 Molybdenum	43 **Tc** 98 Technetium	44 **Ru** 101 Ruthenium	45 **Rh** 103 Rhodium
6	55 **Cs** 133 Cesium	56 **Ba** 137 Barium	57 **La** 139 Lanthanum	72 **Hf** 178 Hafnium	73 **Ta** 181 Tantalum	74 **W** 184 Tungsten	75 **Re** 186 Rhenium	76 **Os** 190 Osmium	77 **Ir** 192 Iridium
7	87 **Fr** 223 Francium	88 **Ra** 226 Radium	89 **Ac** 227 Actinium	104 **Rf** 261 Rutherfordium	105 **Db** 262 Dubnium	106 **Sg** 266 Seaborgium	107 **Bh** 264 Bohrium	108 **Hs** 277 Hassium	109 **Mt** 268 Meitnerium

Lanthanide Series

58 **Ce** 140 Cerium	59 **Pr** 141 Praseodymium	60 **Nd** 144 Neodymium	61 **Pm** 145 Promethium	62 **Sm** 150 Samarium	63 **Eu** 152 Europium	64 **Gd** 157 Gadolinium

Actinide Series

90 **Th** 232 Thorium	91 **Pa** 231 Protactinium	92 **U** 238 Uranium	93 **Np** 237 Neptunium	94 **Pu** 244 Plutonium	95 **Am** 243 Americium	96 **Cm** 247 Curium

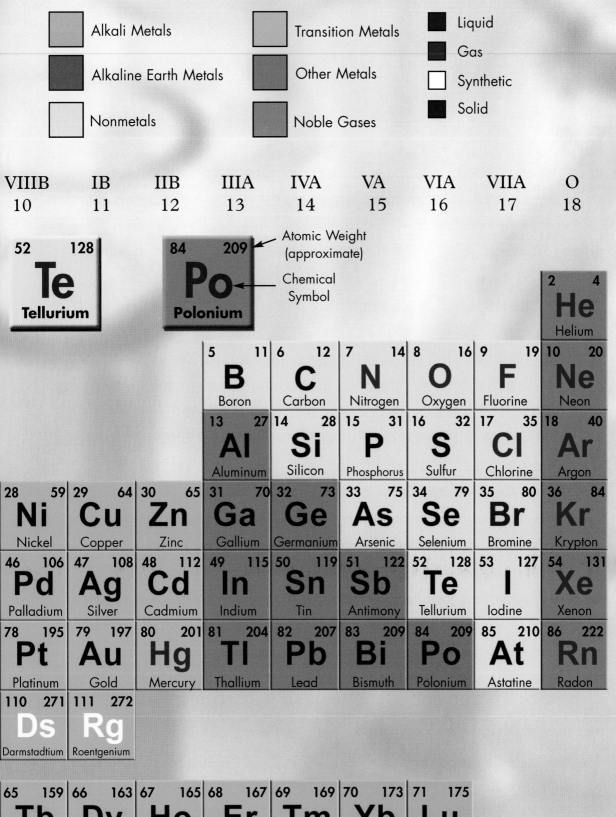

Glossary

alloy A substance composed of two or more metals, or of a metal and nonmetal, that are fused together; often, the components dissolve together when melted.

atom A basic unit of matter consisting of a dense, central, positively charged nucleus surrounded by a cloud of electrons.

atomic number The number of protons found in the nucleus of an atom.

atomic weight The mass of an average atom of an element.

chemical element A substance that cannot be separated into other substances. It is composed of only one type of atom.

compound A substance consisting of two or more different elements bonded together in a fixed ratio.

electron A subatomic particle that carries a negative electric charge.

isotope Any of two or more forms of an element having different atomic masses. The atoms have the same number of protons but different numbers of neutrons.

matter Anything that has both mass and volume or takes up space.

metalloid An element that has some properties of typical metals and some properties of nonmetals.

neutron The uncharged, or neutral, particle in the nucleus of an atom.

nonmetallic An element that has no metal or metal-like traits.

nucleus The central portion of an atom that contains most of the atom's mass.

property A quality or trait that is specific to a certain element.

proton The positively charged particle in the nucleus of an atom.

American Association for the Advancement of Science (AAAS)

1200 New York Avenue NW

Washington, DC 20005

(202) 326-6400

Web site: http://www.aaas.org

The AAAS is a nonprofit organization dedicated to advancing science around the world by serving as an educator, leader, spokesperson, and professional association.

American Association for Clinical Chemistry (AACC)

1850 K Street NW, Suite 625

Washington, DC 20006-2215

(800) 892-1400

Web site: http://www.aacc.org

The AACC is a society of clinical laboratory professionals, physicians, research scientists, and other individuals involved with clinical chemistry and other clinical laboratory science-related disciplines.

American Chemical Society (ACS)

1155 Sixteenth Street NW

Washington, DC 20036

(800) 227-5558

Web site: http://www.acs.org

Founded in 1876, the ACS is the world's largest scientific society. It is a resource for chemistry and chemical applications, including databases, publications, career information, and educational resources.

Canadian Society for Chemistry (CSC)
130 Slater Street, Suite 550
Ottawa, ON K1P 6E2
Canada
(613) 232-6252
Web site: http://www.cheminst.ca
The CSC is the national technical association representing the field of chemistry and the interests of chemists in industry, academia, and government.

Chemical Heritage Foundation (CHF)
315 Chestnut Street
Philadelphia, PA 19106
(215) 925-2222
Web site: http://www.chemheritage.org
The CHF serves the community of the chemical and molecular sciences—and the wider public—by treasuring the past, educating the present, and inspiring the future.

Web Sites

Due to the changing nature of Internet links, Rosen Publishing has developed an online list of Web sites related to the subject of this book. This site is updated regularly. Please use this link to access this list:

http://www.rosenlinks.com/uept/oxel

For Further Reading

Conley, Kate A. *Joseph Priestley and the Discovery of Oxygen* (Uncharted, Unexplored, and Unexplained). Hockessin, DE: Mitchell Lane Publishers, 2005.

Curran, Greg. *Homework Helpers: Chemistry*. Franklin Lakes, NJ: Career Press, 2004.

Hill, James C. *Chemistry: The Central Science* (Student's Guide). Upper Saddle River, NJ: Prentice Hall, 2009.

Newmark, Ann. *Chemistry*. New York, NY: DK Children, 2005.

Saunders, Nigel. *Oxygen and the Group 6 Elements*. Portsmouth, NH: Heinemann, 2003.

Timberlake, Karen C. *General, Organic, and Biological Chemistry: Structures of Life*. Upper Saddle River, NJ: Prentice Hall, 2009.

Tocci, Salvatore. *Oxygen* (True Books). New York, NY: Children's Press, 2005.

Bibliography

Beckett, M. A., and A. W. G. Platt. *The Periodic Table at a Glance*. Oxford, England: Blackwell Publishing, 2006.

Bentor, Yinon. "Oxygen." ChemicalElements.com. Retrieved February 1, 2009 (http://www.chemicalelements.com/elements/o.html).

Bentor, Yinon. "Polonium." ChemicalElements.com. Retrieved February 1, 2009 (http://www.chemicalelements.com/elements/po.html).

Bentor, Yinon. "Selenium." ChemicalElements.com. Retrieved February 1, 2009 (http://www.chemicalelements.com/elements/se.html).

Bentor, Yinon. "Sulfur." ChemicalElements.com. Retrieved February 1, 2009 (http://www.chemicalelements.com/elements/s.html).

Bentor, Yinon. "Tellurium." ChemicalElements.com. Retrieved February 1, 2009 (http://www.chemicalelements.com/elements/te.html).

Bren, Linda. "Oxygen Bars: Is a Breath of Fresh Air Worth It?" FDA.gov, December 2002. Retrieved February 1, 2009 (http://www.fda.gov/Fdac/features/2002/602_air.html).

Brock, William H. *The Chemical Tree: A History of Chemistry*. New York, NY: W. W. Norton and Co., 1992.

EnvironmentalChemistry.com. "Periodic Table Elements: Oxygen." February 22, 2007. Retrieved February 2, 2009 (http://environmentalchemistry.com/yogi/periodic/O.html).

Lane, Nick. *Oxygen: The Molecule That Made the World*. New York, NY: Oxford University Press, 2004.

Loren, Karl. "The Science of Oxygen." OxygenTimeRelease.com, May 20, 2008. Retrieved February 2, 2009 (http://www.oxygentimerelease.com/A/ScienceOxygen/index.htm).

Research School of Biological Sciences. "Discovering Oxygen." Retrieved February 2, 2009 (http://www.rsbs.anu.edu.au/O2/O2_4_Discovery.htm).

Rouvray, Dennis H., and R. Bruce King, eds. *The Periodic Table: Into the 21st Century*. Baldock, England: Research Studies Press, 2004.

Smith, S. E. "What Is the Difference Between Air and Oxygen?" WiseGeek.com. Retrieved February 2, 2009 (http://www.wisegeek.com/what-is-the-difference-between-air-and-oxygen.htm).

Stwertka, Albert. *A Guide to the Elements*. New York, NY: Oxford University Press, 2002.

Index

About the Author

Laura La Bella is the author of ten nonfiction books, many of them focusing on scientific topics and the relevance of chemistry to human and environmental health. Most recently, she has reported on the declining availability of the world's freshwater supply in *Not Enough to Drink: Pollution, Drought, and Tainted Water Supplies* and has examined the food industry in *Safety and the Food Supply*. La Bella and her husband live and work in Rochester, New York.

Photo Credits

Cover, pp. 1, 15, 17, 40–41 by Tahara Anderson; p. 4 © www.istockphoto.com/ Kerrie Kerr; p. 7 © Kenneth Ewald/Photo Researchers; p. 9 © Marilyn Chillmaid/ Photo Researchers; p. 11 © Mary Evans/Photo Researchers; p. 12 © Dr. George Chapman/Visuals Unlimited; p. 21 © www.istockphoto.com/George Clark; p. 22 © www.istockphoto.com/Mikhail Kondrashov; p. 23 © Time & Life Pictures/Getty Images; p. 24 © www.istockphoto.com/Jeffrey Hayden-Kaye; p. 26 © Theodore Gray/Visuals Unlimited; p. 29 © www.istockphoto.com/Achim Prill; p. 32 © Visuals Unlimited; p. 33 © www.istockphoto.com; pp. 34, 35 NASA; p. 36 © www.istockphoto.com/ Bryan Faust; p. 38 © www.istockphoto.com.

Designer: Tahara Anderson; Photo Researcher: Marty Levick